Smelly Old History

MOULDY MUMMIES

Mary Dobson

OXFORD UNIVERSITY PRESS

Oxford University Press, Great Clarendon Street, Oxford OX2 6DP

Oxford New York
Athens Auckland Bangkok Bogotá Bombay
Buenos Aires Calcutta Cape Town Dar es Salaam
Delhi Florence Hong Kong Istanbul Karachi
Kuala Lumpur Madras Madrid Melbourne
Mexico City Nairobi Paris Singapore
Taipei Tokyo Toronto Warsaw

and associated companies in
Berlin Ibadan

Oxford is a trade mark of Oxford University Press

© Mary Dobson 1998
First published 1998

Artwork: Chris Smedley. Photographs: British Museum, London: 8, 19, 26, 27; Professor
G. T. Martin: 15.

A CIP catalogue record for this book is available from the British Library

ISBN 0-19-910495 6

1 3 5 7 9 10 8 6 4 2

Printed in Great Britain

Contents

A Sense of the Past 4

Smells Divine 6

Deadly Deodorants 8

Spicy Stuffings 10

Foul Pharaohs 12

Meaty Mummies 14

Fragrant Feasts 16

Lotions and Potions 18

Egyptian Effluence 20

Pyramid Problems 22

Brutal and Barmy 24

Aromatic Animals 26

A Bed of Roses 28

Pungent Puzzles 30

Glossary 31

Index 32

Scratch the scented panels lightly with
a fingernail to release their smell.

A SENSE OF THE PAST

The ancient Egyptians stuffed their bodies - alive and dead - with spicy scents. They weren't barmy! It all made perfect sense to them: while alive they could conceal their awful odours, and once dead they could preserve their mummies (and daddies) for ever. Some of these embalmed mummies have survived for thousands of years; sadly, others went mouldy rather quickly.

Of all the senses of the past, we often forget the sense of smell! This book takes you as close as possible to smelly old history. We've sniffed out some wonderful ancient aromas, as well as some pretty foul ones, and we've opened up a nice selection of mouldy and meaty mummies for you to sample!

The ancient Egyptians who lived along the River Nile 5,000 years ago fished and farmed, and did all the usual basic things in life. But they also came up with some amazing arrangements for a heavenly after-life. They built magnificent pyramids for their freshly festering pharaohs, and they hid their fragrant old remains in deep tombs. Some of their secrets lay undisturbed for centuries. Then, historians poked their noses into the past and began to unwrap the story...

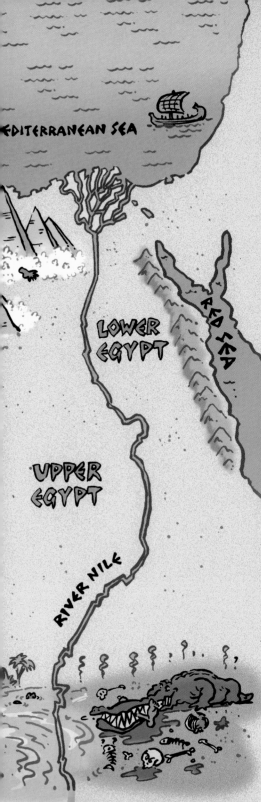

MOULDY MUMMIES

In ancient Egypt's glorious past
The pharaohs tried to make life last.
They stuffed the dead with herbs and spice
To keep thier bodies smelling nice.
At first their scheme went badly wrong,
And mouldy mummies caused a pong!

But then they found a clever trick
To stop the ancients looking sick.
They popped their bodies on a bench
And took out bits that caused a stench:
The brains, the guts and all that rot
Were carefully stored inside a pot.

With scented wraps and empty tummies
They made some lovely smelly mummies.
Their pyramids were built up high
To waft their souls into the sky.
The mummies lay at peace inside,
Until grave robbers poked and pried.

But mortal life was not all sweet,
And even pharaohs had smelly feet.
For those who lived along the Nile
The crocs and pongs were pretty vile.
Just hold your nose and take a look
At all the whiffs inside this book!

SMELLS DIVINE

Trying to stop everything going mouldy in the hot desert sun was no joke. But the ancient Egyptians quickly sensed that strong perfumes, like myrrh and frankincense, could overpower foul stenches and also please their gods. Nowhere smelt more heavenly than the temples. Sweet-smelling incense was burnt morning, noon and night. Even the statues were scented. During the 30 years' reign of Rameses III, 1,933,766 jars of incense, honey and oil were used in the temple of the god Amun-Re.

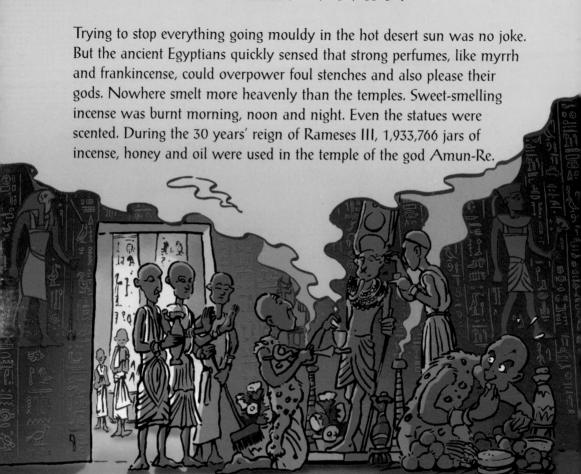

This priest is exceptionally pure. He has already washed twice during the night in a sacred pool, and shaved his body all over to remove lice and other impurities. He is refreshing the gods with incense, which he believes is their own sweat fallen to earth.

His friend is nibbling a few of the gods' perfumed leftovers. They taste better than the cedar gum he usually chews to keep his breath fresh.

The gods were the purest of all and had a wonderful fragrance. They came in all shapes and sizes, often with animal heads.

Osiris, lord of the underworld.

Isis, his beautiful wife.

Horus, their son, protector of the pharaoh.

Nefertem, lord of the sweet smells.

DEADLY DEODORANTS

Egyptians wanted to preserve their bodies after death so that their spirits - Ka and Ba - had a home in the after-life. If they looked lifelike, the spirits would easily recognise their own mummies.

Mummy matters in the very early days were dead easy – no fuss and no mess. The dead body was simply placed in the sand – well away from any living bodies - and left to dry out in the blazing hot sun. Some got left out rather a long time!

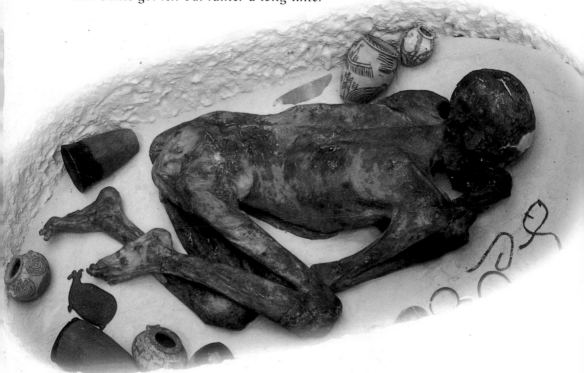

This delightful little chap, nicknamed Ginger, lay curled up in the desert for over 5,000 years before moving indoors to the British Museum in London.

Then somebody had the idea of putting the corpses in tombs. You might think this was a brilliant idea, but it had deadly consequences. The body no longer dried out in the sun - instead it went rotten and mouldy. The smell was so bad that some other bright spark decided to wrap the corpse up in bandages and dose it with deadly deodorants, before sticking it in the tomb. This was even worse, and mouldy mummies became a grave issue.

While some nobles pop a fresh body into its tomb, Neferhotep has sneaked off to do a spot of grave-robbing. He gets a shock when he opens this one! It's a crocodile!

Scratch and sniff for a mouldy whiff.

SPICY STUFFINGS

Finally, the Egyptians came up with a new solution to stop bodies going mouldy. They packed them with a salty substance called natron, to soak up the body fluids, then stuffed them with spices to keep them fresh and fragrant for ever. Making mummies took place in an embalmer's workshop called 'The Pure Place', but it was a pretty messy business!

The embalmer removes all the smelly insides like the guts, the bowels and the lungs. The brain is pulled out through the nostrils with a bronze hook. These offensive organs are put into canopic jars.

The body is now washed, packed with natron and left for 40 days to dry out, while the embalmer takes a breather.

This mummy is having the full works - stuffed and spiced with sawdust, fragrant herbs, oils and perfumes. An old wound is patched up with gazelle skin, a missing leg is replaced with a wooden one, and a couple of onions are popped into her eye sockets.

Her body is wrapped up with 20 layers of fresh linen bandages, which are coated with a sticky resin to glue them together. Sadly, some relatives couldn't afford all the spicy stuffings for their loved ones, so the mummy just got a few mouldy rags.

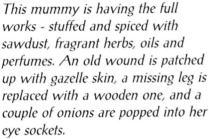

Finally, her mouth is opened so she can eat and drink in the after-life, if she gets that far. This will be decided by the gods in the Hall of Two Truths. Anubis, the jackal-headed god of mummification, weighs her heart against all her bad behaviour. Too bad, this one's off to the Devourer of the Dead!

FOUL PHARAOHS

The pharaohs were the great rulers of Egypt. They were thought to be gods ('pharaoh' means 'great house'). Tutankhamun was only nine when he became king, and about 18 when he died in 1327 BC. Visitors were expected to kiss the ground beneath his feet - but beware, the Caretaker of the Royal Toe-Nails has forgotten to put any perfume on his smelly little tootsies.

Scratch and sniff for a cheesey whiff!

Pharaohs' mummies were given richly fragrant farewells. Their multi-layered coffins, decked with flowers, perfumes, gold and other useful goods (like model workers) for the after-life, were placed in barges and taken up the River Nile to magnificent pyramids, or stone-cut tombs, in the desert. These were designed to hide their burial chambers, and to keep robbers off the scent. But this didn't always work.

In 1922 Tut's tomb was discovered by Howard Carter in the Valley of the Kings. Its reeking ruins had already been disturbed a few thousand years earlier: grave robbers had emptied all the jars of perfume on to the floor - one had even left his fingerprint inside a jar as he scooped out the last dregs! But Tut's mummy was still in one piece - or just about. The embalmers had used so many spices that when it was unwrapped, his little finger fell off.

MEATY MUMMIES

Opening up ancient Egypt has been an exciting - and smelly - business! One archaeologist, who poked his nose into a tomb, got the full whiff of smelly old history. He later wrote: 'I closed my eyes and I smelt incense ... I smelt time ... I smelt centuries ... I smelt history itself.'

Others had a less fruitful experience: in 1798 the French Emperor, Napoleon, went alone into the Great Pyramid of Cheops. He came out deathly pale and gasping for air - he never told anyone what he had seen and smelt.

The word 'mummy' actually comes from 'mumiya', an Arabic word for bitumen or tar. The Arabs who invaded Egypt in the 7th century thought the ancient bodies were coated in bitumen, and called them mummies.

From the middle ages, explorers began to pull bits off the mummies to take home as souvenirs. Mummy skin, mummy paste and mummy powder were especially popular as medicines!

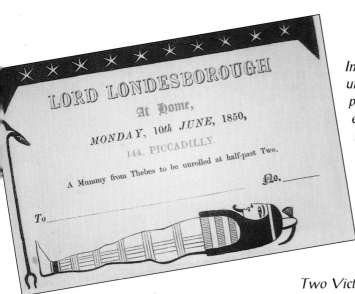

In the 19th century, unwrapping mummies in public was a popular event: this invitation to a mummy party was accepted by a large number of nosy onlookers!

Two Victorian ladies who tried to smuggle a mummy out of Egypt were caught red-handed - the smell gave them away! This smell became known as the 'mummy bouquet'.

Peppercorns and tiny animal bones were found stuffed up the nose of Rameses II to keep it upright. He was about 90 years old when he had his nose job done by the embalmers.

Horemkensei, a priest, was the last mummy to be unwrapped in Britain, when he started to go a bit mouldy in the hot summer of 1976. Now special scanners can view the innards without unwrapping the outers.

FRAGRANT FEASTS

The Egyptians weren't just hoping for a good death - they were out to enjoy life too! This balmy banquet is overflowing with gold, frankincense and myrrh. The guests are wearing the latest head-gear: a lump of animal fat steeped in scent. The table groans under the weight of fresh flowers. Even the food - served on golden platters - reeks of spicy scents and pungent perfumes.

Khety, the servant, greets a guest at the entrance. She offers him his favourite deodorant.

Meryt and Paneb are invited with their mummies and daddies. They have brought a toy crocodile to keep them amused, while the grown-ups gorge themselves on aromatic antelope, greasy gazelle and stuffed goose livers.

Ahmose's perfumed cone is a bit overpowering. Fortunately, the nasty yellow stain on her linen robe will wash out in the Nile.

Scratch and sniff for a frankincense whiff.

Poor peasants had a much less fragrant diet and many had to put up with mouldy bread, garlic and bad breath.

LOTIONS AND POTIONS

Keeping the skin soft, silky and scented was an essential part of life.
Beauty tips and doctors' orders were written down in hieroglyphs on
papyrus scrolls (a type of paper made from reeds in the Nile). It is only in
the last 200 years that historians have been able to decipher these
delightful and disgusting recipes.

'RUB YOUR BODY ALL OVER WITH OIL OF KYPHI – IT WILL
SMOOTH YOUR CURVES AND SOOTHE YOUR NERVES. FOR A
CHEAPER SOLUTION, TRY HIPPOPOTAMUS FAT.'

'PAINT YOUR EYES REGULARLY WITH KOHL.
YOU WILL LOOK STUNNING, AND YOUR SORE
EYES WILL BE PROTECTED FROM SUN, DUST
AND FLIES.'

'TO RESTORE LOSS OF EYESIGHT, POUR
HONEY, RED EARTH AND A GROUND-UP
PIG'S EYE INTO YOUR EAR. RECITE A
SPELL AT THE SAME TIME.'

'GREY HAIR IS AN UNNECESSARY SHAME. TRY RUBBING ON BLACK COW BLOOD, BLACK SNAKE FAT AND RAVENS' EGGS TO TURN IT BACK TO BLACK.'

If that doesn't work, try a freshly made wig like this one.

'MOULDY BREAD AND MEATY MAGGOTS WILL MOP UP AN OOZING WOUND IN NO TIME.'

'TO GET RID OF EVIL SPIRITS, USE ONLY THE FOULEST INGREDIENTS: LIZARD FILTH, GAZELLE DUNG, FISH BILE, CRUSHED COW'S LIVER AND THINGS A CROCODILE DIGESTS ARE ALL GUARANTEED TO REPEL EVEN THE MOST REVOLTING SPIRITS.

Can you believe that these rotten remedies actually worked! We now know that mouldy bread contains antibiotics which kill bugs, while meaty maggots do a good job licking up old sores. Honey is one of the oldest natural remedies, and is still used today to heal wounds.

EGYPTIAN EFFLUENCE

Rich Egyptians certainly knew how to look after their scented bodies. But in spite of their affluence, they were much less careful about their effluence. Waste products were thrown on to the streets and into the river. Most settlements, like this one, were built close to the River Nile - the water was vital for drinking, washing, and watering crops. But, sadly, its smelly contents contained nasty germs which wormed their way through Egyptian society.

Ameny is rushing inside to use the latrine. A servant deals with the effluence which gushes along the town gully and is quickly recycled.

These children are enjoying a game on the river - Meri-teti doesn't notice the tiny worms in the water which will go in one end and out the other, causing terrible stomach upsets.

Most of the houses are plagued by rats. Fortunately, there is a good old-fashioned Egyptian remedy: cat's grease and gazelle dung have been smeared round this house to keep the rats away.

Scratch and sniff for a repellent whiff.

PYRAMID PROBLEMS

The pyramids are some of the greatest wonders of the ancient world. The first pyramid was built in 2680 BC. It had steps to take the pharaoh up to the sun-god. The largest was the Great Pyramid at Giza, built in about 2500 BC for King Khufu. Its base is the size of eight football pitches, and it is made from over 2 million limestone blocks. The pharaoh kindly gave his loyal subjects time off work to build it! It took thousands of men and 20 years' hard labour to complete. Imagine the sweat and toil!

Some of the later pharaohs, like Tut, must have thought the pyramids were a bit over the top. They chose a nice shady spot deep in the cliffs of the Valley of the Kings for their resting place. The tomb builders lived in cramped, smelly huts, but were richly rewarded with a daily ration of radishes, garlic, onions, massaging oils, linen and bread.

Some were a bit rebellious, and found amazing excuses for having the odd day off. One chap said he'd been stung by a scorpion. Another had problems sorting out his aunt's mummy. One had to take his donkey to the vet. The final crunch came in 1170 BC, when supplies were 20 days overdue. The workers revolted (they were already a bit revolting, but this was too much). They went on strike until they were given their onions. This is the first recorded strike in history!

BRUTAL AND BARMY

Egyptians worked pretty hard building pyramids, mining gold, keeping their pharaohs happy, and stopping their mummies from going mouldy. Fancy trying some of these odorous occupations?

Embalmer
This is a dead easy job if you don't mind being up to your elbows in grease.

Ripper-upper
This job comes with no strings attached - it's just a matter of knowing where to pull, to get the brains and guts out!

Guardian of the royal bowel movement
This richly rewarding position is suitable for anyone who cares about the pharaoh's tender needs.

Common barber
Try your hand at shaving, scalping, delousing and deodorising. Don't forget to pick up the nits and bits of old hair afterwards - you can stick them together with beeswax to make freshly scented wigs.

Daytime gold-miner and night time grave-robber
A golden opportunity for those who prefer a shady deal. But beware of the spooky stench and the pharaohs' curse: some say that those who go in never come out . . .

Slave
This job has no rewards except brutal treatment and sweaty service. Only on offer to criminals and prisoners of war.

25

AROMATIC ANIMALS

The Egyptians gave their favourite pets fragrant funerals. When a cat died, everyone in the household shaved their eyebrows in respect. Cats were mummified like humans, then popped into cat coffins. If you hadn't already got one, you could buy a mummy cat outside the temple.

Insects, like the scarab beetle, were also popular. The scarab rolled balls of dung along the ground, setting a good example on how to clean up! Many mummies wore aromatic scarab-shaped amulets to ward off foul spirits.

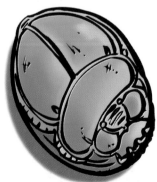

37348

Some creatures weren't so friendly, but they still got royal treatment. The crocodile was one of the most feared reptiles, so the Egyptians came up with the bright idea of making it sacred. That way anyone who was eaten by a croc also became sacred - not much consolation for a grisly end!

Hippos were a mixed blessing. Taweret the pregnant hippo goddess looked after mothers and babies, but other hippos were too large for comfort. This hippo is wallowing in the mud, while a few desperate fishermen are trying to get past.

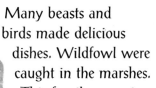

Many beasts and birds made delicious dishes. Wildfowl were caught in the marshes. This family are using a boomerang, or throw-stick, to catch their prey. Their cat is also enjoying the family outing!

A BED OF ROSES

For over 2000 years, a succession of foul and fragrant pharaohs came and went. In the last 1000 years BC, foreigners began to invade Egypt, including the Persians and the Greeks. Cleopatra came from a family of Greek rulers in Egypt. She was married to her brother, Ptolemy, but they often squabbled. Cleopatra rather fancied herself, and she decided to get the Romans to help her overpower her brother. She overpowered the Romans with her heavenly smell (and great charm), but she came to a tragic end.

First, she captivated Julius Caesar (she rolled up in front of him in a rug!). When he'd been murdered, she captured the heart of Mark Antony, another Roman general.

Sadly, Cleopatra's rosy dreams didn't last long. In 30 BC Mark Antony heard a rumour that she was dead, and he killed himself. When Cleopatra heard of this, she too killed herself in despair.

After that, the Romans took over Egypt. They even tried to make mummies, but they did a rotten job. They forgot to take the insides out, and they stuck bits in the wrong place. Finally, when the Arabs took over Egypt in the 7th century AD, making mouldy mummies was abandoned for good.

Scratch and sniff this fragrant scene.

Cleopatra's silky skin is drenched in the finest oils, her barge is awash with aromatics - even the sails are scented - and dripping with roses.

PUNGENT PUZZLES

In the past, mummy parties were in fashion; today, ancient mummy jokes are in fashion - you must have heard about:

- the ancient Egyptian boy who was confused because his daddy was a mummy

- the knocker on the mummy's tomb which said 'Toot and Come In'

Make up your own mouldy mummy jokes.

Spell the smell
Imagine how the pharaohs smelt! Untangle these words to find out:

 WOLFERY

 GREETFINS

 ULOF

 GNATFRAR

 NATSLEASINO

 LUMYOD

 FREEPMUD

 TWIH-RITEH-SESON

GLOSSARY

Amun-Re	Amun was the chief god of Egypt. Re was the sun god.
amulet	An object worn as a charm against evil spirits.
Anubis	The jackal-headed god of embalming.
archaeologist	A person who digs up and studies historical remains.
BC	Before Christ. Used to describe the years before the date of Christ's birth. The ancient Egyptians would not have used this term.
canopic jars	Jars used to hold the organs of the dead before embalming.
embalmer	A person who preserves dead bodies.
hieroglyphics	The Egyptian system of writing using picture symbols.
Horus	The god of the sun and sky, with a magic eye.
incense	A substance that burns with a sweet smell.
Isis	The goddess of fertility.
kohl	A cosmetic used for eyes.
kyphi	A richly perfumed oil.
latrine	An ancient lavatory.
mummy	The body of a dead person or animal which has been dried out and preserved.
natron	A chemical used to preserve dead bodies.
Osiris	The god of the land of the dead.
papyrus	A reed-like plant used to make a type of paper.
pharaoh	An ancient Egyptian king.
pyramid	A massive monument with four triangular sides, built as a tomb for Egyptian kings and queens.
Taweret	The goddess of birth.
Valley of the Kings	A narrow valley in Egypt with tombs of the pharaohs.

INDEX

animals   23, 26-27
Arabs   14, 29

cats   26, 27
children   16, 20
Cleopatra   28-29
crocodiles   5, 9, 27

disease   20

embalmer   10-11, 13, 24

farming   4, 27
fishing   4, 27
food   6, 16-17, 18-19, 22, 23
funerals   13

gods   6, 7, 11, 27
Greeks   28

hieroglyphs   18
Horemkensei   15

Julius   Caesar   29

Khufu   22

Mark   Antony   29
medicine   14, 18-19
mummies   4, 5, 6, 7, 8, 9, 10-11, 14-15, 24, 25, 26, 29

Napoleon   14
Nile   4, 5, 13, 17, 18, 20-21

perfumes   4, 6, 13, 16-17, 18, 28-29
Persians   28
pharaohs   4, 5, 12-13, 22, 24, 28
priests   6-7, 15
Ptolemy   28
pyramids   4, 5, 13, 14, 22-23

Rameses II   15
Rameses III   6
religion   6-7
Romans   28, 29

settlements   20-21
slaves   25

temples   6
tombs   4, 9, 22, 25
Tutenkhamun   12-13, 22